MEERKATS

by Golriz Golkar

abdobooks.com

Published by Pop!, a division of ABDO, PO Box 398166, Minneapolis, Minnesota 55439.
Copyright © 2022 by POP, LLC. International copyrights reserved in all countries.
No part of this book may be reproduced in any form without written permission from the publisher.
Pop!™ is a trademark and logo of POP, LLC.

Photos: iStockphoto, cover, 1, 3 br, 4-5, 11, 15, 21 bl, 23, 24 tl;
Shutterstock Images, 3 top, 3 bl, 9, 10, 13, 16, 21 top, 21 br, 24 tr, 24 cl, 24 cr, 24 bl, 24 br;
Illustrations: Heather Bazata, inside front cover, 6, 18

© 2025 CFA Properties, Inc. Chick-fil-A® and the Chick-fil-A Play Logo™ are trademarks of
CFA Properties, Inc. www.chick-fil-a.com Chick-fil-A, Inc., Atlanta, GA 30349
Adapted from the original for Chick-fil-A, Inc. by Frederic Thomas USA, Inc.
Printed in USA, Frederic Thomas USA, Inc., 12/2024 ABDO Animal Books - Meerkats

Marvelous Meerkats

Meerkats are mammals with gray and brown fur. They have pointed faces. Black patches circle their eyes. Their bodies are long. Their tails have dark tips.

Meerkats live in groups called mobs that can have up to 50 meerkats. Several families make up a mob.

Meerkats play, hunt, and fight off predators together.

Meerkats are good fighters. A mob can scare off large animals such as jackals.

We take care of each other and play together.

What kinds of activities do you like to do with your friends and family?

Living in the Desert

Meerkats live in the deserts of southwestern Africa. Their underground burrows keep them cool and hidden from predators.

Meerkats dig with their long, powerful claws. Their tiny ears close to keep dirt out.

Meerkats have an extra set of eyelids. They are thin and clear. When meerkats blink, the eyelids get rid of any sand in their eyes.

Black eye patches reduce the sunlight's glare. Meerkats have large pupils that give them a wider field of vision. When meerkats stand to check their surrounding, they use their tails for balance.

Meerkat Mealtime

Meerkats use their strong sense of smell as they dig for their food. They eat insects and spiders. Meerkats also catch small lizards and birds.

One meerkat is the lookout for predators as others search for food. It stands tall on its back feet. If it senses danger, it warns the others to hide. Meerkats take turns being the lookout.

> Meerkats make many sounds. They can growl, whistle, cluck, and bark.

Let's play Meerkat Peek!

1. One person, the counter, shuts eyes and counts to 10
2. Everyone else hides
3. At 10, counter opens eyes
4. Everyone peeks up then hides
5. First one spotted is the next counter

The Life of a Meerkat

Female meerkats can have two to four babies each year. These pups are born in burrows. After about four weeks, the pups go above ground for the first time.

The whole family teaches pups how to hunt, hide, and stay clean. Meerkats start their own families when they are approximately one year old. Meerkats live up to eight years in the wild.

Creature Quest

Find these six meerkats in this book!

Family fun at your fingertips

Free fun for the whole fam
Enjoy ebooks, shows, games, crafts, and more on the Chick-fil-A Play™ App

©2025 CFA Properties, Inc. All trademarks shown are the property of CFA Properties, Inc.

ABDOBOOKS.COM

THIS SERIES FOLLOWS MABEL AS SHE TRAVELS THE WORLD LEARNING NEW RECIPES TO MAKE AT HOME AND SHARE WITH FRIENDS!

Available for purchase wherever books are sold.

VIEW ALL 6 TITLES AT ABDOBOOKS.COM

Certified Sourcing

www.forests.org
SFI-01060

The fiber in this paper product
is from responsible and legal sources.

Good for you. Good for our forests.®

Hi, I am a meerkat!

Can you give me a name? Then find my meerkat friends in this book. You can color us if you wish. Let's have fun together!

This meerkat's name is: _____

ABDO Animal Books

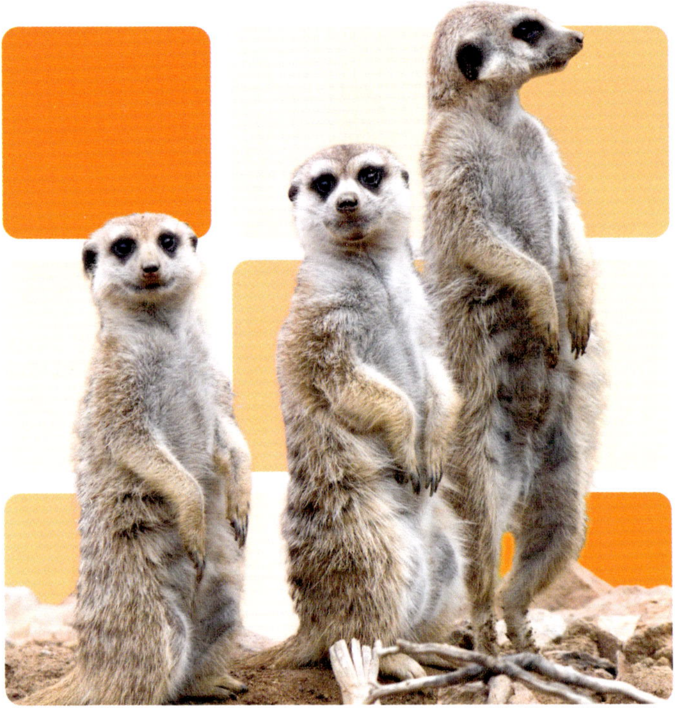

MEERKATS

by Golriz Golkar